◆ Primary Theme Series ◆

Plants

AF855026

Written by Julie Snyder and Kimberlee Graves

Editor: Wendy L. Blocher

Illustrator: Catherine Yuh

Project Director: Carolea Williams

Table of Contents

Introduction		3
Getting Started	Classroom Environment	4
	Management ◆ Assessment	6
	Performance Evaluation (reproducible)	7
	Parent Letter (reproducible)	8
Reading & Language Arts	Blooming Good Books ◆ Flower Legends	9
	Blooming Books (reproducible)	10
	Plant Bookmarks (reproducible)	11
	Seed to Flower ◆ Blooming Personalities	12
	Writing about Plants	13
	Flower Pattern (reproducible)	14
Mathematics	Word Problems ◆ Seed Sort ◆ Popcorn-Kernel Estimating	15
	Sizing Up Plants ◆ Concentrating on Plants ◆ Add a Plant	16
	Blooming Math (reproducible)	17
Science	Exploring Plants Experiments	18
	Plants Experiment Record (reproducible)	19
	Peek-and-See ◆ Growing Mobiles	20
	Hungry for Plants ◆ Parts of a Flower	21
	Plants We Eat! (reproducible)	22
	Parts of a Flower (reproducible)	23
Social Studies & Geography	Flowers across America ◆ A More Beautiful Place ◆ Growing with Responsibility	24
Music & Drama	Musical Plants ◆ Garden Gossip	25
Physical Education	Skip to My Garden ◆ Garden Relay	26
Arts & Crafts	Pressed Flowers ◆ Petal Prints ◆ Crystal Gardens	27
	Beautiful Blooms ◆ Garden Scenes ◆ Pleasant Potpourri	28
Cooking	Chocolate Dirt ◆ Sunshine Cookies	29
Special Events	Field Trip ◆ Garden Tours ◆ Flower Festival ◆ Garden Careers	30
Student Awards		31
Recommended Books		32

Introduction

Use thematic teaching in the classroom and watch your students thrive in a cooperative, creative learning environment that emphasizes thinking, speaking, reading, and writing. CTP's *Primary Theme Series* helps you organize your time, materials, and classroom space while providing fun, thought-provoking activities that all students will enjoy.

The *Plants* book offers a "garden" of information on plant life. Students explore, investigate, analyze, and discuss such plant-related topics as

- the parts of a flower.
- the life cycle of a plant.
- photosynthesis.
- flowers across America.

Hands-on, integrated activities include

- making plant mobiles.
- designing a garden mural.
- graphing plant growth.
- making seed mosaics.
- cooking up chocolate "dirt."

The *Plants* book also provides several reproducible sheets, including a performance-evaluation form, a parent letter, and student certificates. Your students will "bloom" with excitement as they explore a garden of plants.

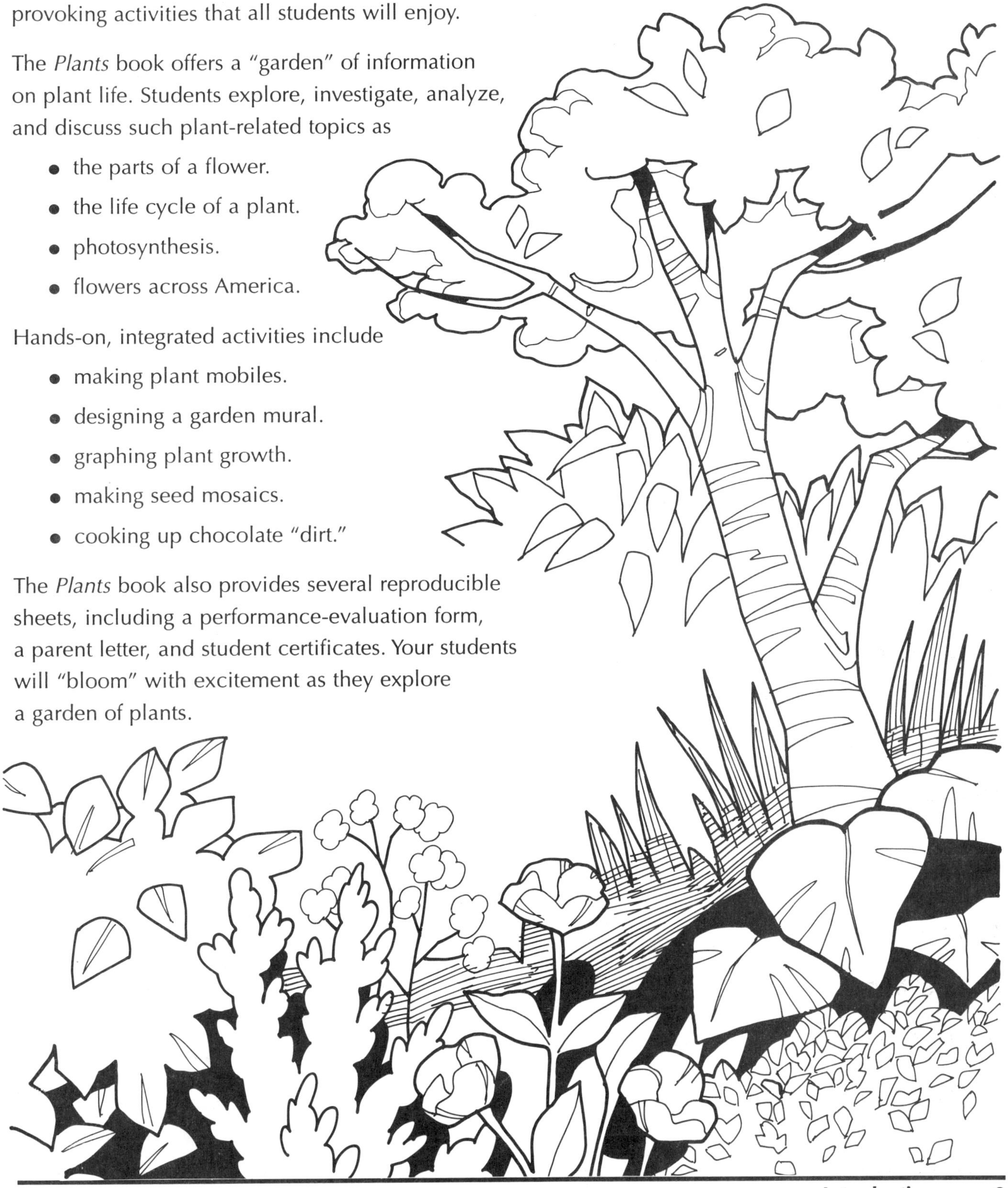

Getting Started

Classroom Environment

Set up colorful displays and learning centers children can freely use to observe and learn about plants. Set up a display or science table that includes several different types of seeds and plants. Provide soil and planters for students to use when planting seeds. Set up an inviting reading area that includes a class library of books relating to plants (see Recommended Books, page 32). For extra fun, store gardening tools, pens, and pencils in clay pots decorated with paint and stickers.

◆ ◆ ◆ ◆ ◆ ◆ ◆ ◆ ◆ ◆ ◆ ◆ ◆

Garden Time Mural

Invite students to create a wall mural titled *Garden Time*. Cover a large bulletin board or wall area with blue butcher paper. Glue brown butcher paper to the bottom of the display to create "soil." Provide a variety of craft supplies (e.g., paint, construction paper, cotton, fabric scraps, felt, yarn, glitter, feathers, googly eyes, and tissue paper), and have students use the following suggestions to make the "garden":

- Create a 3-D sky by gluing stretched-out cotton to the top of the blue butcher paper. Add a yellow construction-paper sun. Add glitter to the sun for a sparkling effect.
- Make colorful plants and flowers out of construction and tissue paper, and arrange them in rows on the soil. Use burlap fabric to make dirt and rocks. Use white construction-paper strips to create a picket fence.
- Make a cheerful scarecrow out of textured fabric scraps.
- Use various craft items to make garden guests, such as rabbits, birds, and insects.

Seed Study Station

Set up planters and soil to simulate a garden. Add gardening tools, a watering can, and other small gardening items. Invite students to donate dried beans, peanuts, sunflower seeds, fruit seeds, nuts, and packaged seeds. Glue seeds to tagboard or index cards, identify each one by name, and add a picture of the full-grown flower or plant. For extra fun, have students sort seeds into empty egg cartons. Provide magnifying glasses so students can easily observe and compare the different sizes, shapes, colors, and parts of the seeds. Encourage students to plant the seeds in planters of soil to make a class garden. Provide garden clothing, such as overalls, straw hats, and garden gloves, for students to use during their seed study or creative play.

◆ ◆ ◆ ◆ ◆ ◆ ◆ ◆ ◆ ◆ ◆ ◆ ◆

Charts

Draw a plant stalk on chart paper. Ask students to list on the chart words that relate to plants, such as *roots, seeds,* and *soil.* Title the chart *Words That Grow,* and post it in the classroom. Throughout the unit, invite students to record on the chart theme-related vocabulary words they have learned. To further challenge students, have them classify the words as nouns, adjectives, verbs, and adverbs.

Also make a KWL chart that includes the column headers *What We* ***K****now about Plants, What We* ***W****ant to Learn about Plants,* and *What We* ***L****earned about Plants.* At the beginning of the unit, have students brainstorm ideas for the first two columns. At the end of the unit, invite students to share what they learned about plants as you record their ideas on the chart. Invite students to review and compare what they knew about plants to what they learned.

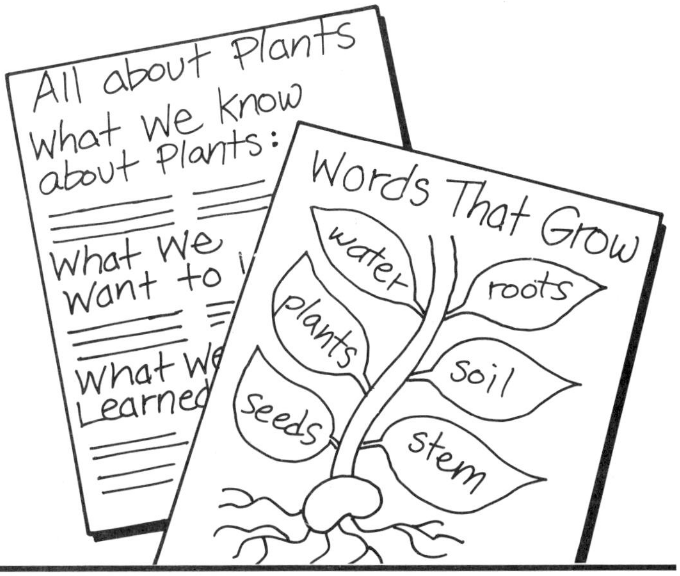

Management

The amount of time you spend on this unit depends upon your goals and objectives. As you teach students about plants, consider the following suggestions:

- Vary instructional strategies so activities validate different learning modalities—auditory, visual, tactile, and kinesthetic. Encourage students to investigate, analyze, speculate about, and verify answers to problems. Follow up verbal instructions with visual cues and hands-on applications.
- Encourage peer interaction. Organize your class into small groups so students can share creative ideas, art supplies, and manipulatives. Use the reproducible certificates (page 31) to recognize and reward your "plant experts."
- Provide learning centers with multilevel, self-paced activities addressing various learning styles. Store learning-center materials in colorful, portable containers next to each corresponding station. Have students place completed paperwork in holding trays, file folders, or work cubbies. Post a laminated class list, and ask students to check off their name when they have completed an activity.
- Encourage family involvement through written correspondence (see the reproducible parent letter, page 8) and take-home activities. Provide books (see Recommended Books, page 32) that students can check out and share with family members.

◆ ◆ ◆ ◆ ◆ ◆ ◆ ◆ ◆ ◆ ◆ ◆ ◆

Assessment

Assessment should be an integrated part of regular instruction. Include a variety of approaches, such as scoring rubrics (see Performance Evaluation, page 7), anecdotal records, and student portfolios. Encourage children to share in the evaluation of their work, selecting and self-evaluating completed assignments to include in their portfolios. (Date all portfolio entries so they can be chronologically compared at any time.) Conduct individual student conferences to verify understanding of concepts as well as to assess the ability to follow directions. Ask students to discuss their most interesting activity, prompting them with questions such as *What did you learn by doing this activity?* or *What might you do differently if you tried this activity again?*

Performance Evaluation

Student ______________________________ Date ______________________________

Check the level that best reflects the student's performance.

Activity: ______________________________	Performance Level			
	Excellent	Very Good	Good	Needs Improvement
Shows motivation and curiosity for learning.				
Demonstrates full understanding of concepts.				
Clearly communicates and listens to others.				
Accurately records and describes observations.				
Uses knowledge to solve problems or extend thinking.				
Comments:				

Rubric

Excellent

Goes beyond competency, adding creativity and insight to overall performance. Shows initiative and takes charge of own learning. Listens attentively to others. Shows advanced critical thinking skills. Written work is polished, with detailed explanations that extend into other subject areas.

Very Good

Uses skills effectively. Listens well during discussions, contributing thoughtful ideas and opinions. Work is neat and accurate, showing evidence of higher-level thinking. Does not take risks or extend ideas into other subject areas.

Good

Shows much effort and desire to learn but is still working on mastery of skills. Written work is accurate but shows little creativity or higher-level thinking. Follows directions well but needs extra encouragement and time to organize work.

Needs Improvement

Lacks organization and effort. Student is unsure of how to use materials or uses them incorrectly. Written work is inaccurate and shows little or no creativity. Does not follow directions and needs additional guidance to perform general tasks.

Dear Parents,

Our class is beginning an exciting unit on plants. We will be reading, writing, investigating, and discussing topics such as the parts of plants, the life cycle of plants, and what plants need to grow.

To help us with our plant investigations, we would appreciate you sending any of the following materials to class with your child:

assortment of seeds	*fabric scraps*	*plastic knives or spoons*
planting soil	*coffee filters*	*empty egg cartons*
planters	*potpourri oil*	*old nature magazines*
waxed paper	*clay or plastic pots*	*spices (cinnamon, allspice, rosemary)*

Share in your child's excitement for learning by completing the following activities together:

- Set aside time each night to read aloud books about plants and flowers, such as *The Tiny Seed* by Eric Carle and *The Carrot Seed* by Ruth Krauss. Be sure to include a variety of stories, including folktales, adventures, and mysteries.
- Take your child on a plant hunt at a local botanical garden, in your neighborhood, or at a nearby park. Bring along a field guide to help you identify different plants.
- Plant a small garden with your child. Have your child keep a growth journal as the plants mature. Invite your child to share his or her observations with classmates.
- Make a family plant book. Have your child and other family members find drawings or magazine pictures of plants. Invite your child to write interesting and unusual facts about the plants.
- Make a garden soup from plant ingredients, such as carrots, broccoli, potatoes, beans, and celery. Serve the soup as part of a yummy family meal.

Thank you for your support.

Sincerely,

Blooming Good Books

Cover a large bulletin board or wall area with colorful butcher paper. Twist green butcher paper into a vine shape, and staple it to the bulletin board. Add paper leaves, buds, and insects. Give each student a Blooming Books sheet (page 10). Encourage students to read a book about plants and write a report about it on their sheet. Have students color their flower, cut it out, and pin it to the bulletin-board vine. To encourage more reading, give each student a flower bookmark (page 11) to use while reading favorite plant books.

Flower Legends

Read aloud *The Legend of the Indian Paintbrush* by Tomie dePaola. After discussing the story with the class, show students how to make their own paintbrush flowers. Roll a colorful sheet of construction paper into a tube, and secure it with tape. Cut slits in one end, and pull them back to create "bristles." Invite each student to make several flowers and a pot to put them in. Have students cut a small portion from the top of a folded paper bag and fold it down. Encourage students to use paint, colored markers, or crayons to add designs to their bag. Then have students stuff their bag with newspaper and place their flowers inside. Invite students to write their own legend about a real or an imaginary flower's origin and then share their story and flowerpot with the class. Use the students' creations as part of a bulletin-board display for everyone to enjoy.

Blooming Books

Directions: Fill out the information below. Then cut out the flower. Roll the petals of the flower around a pencil so they will curl.

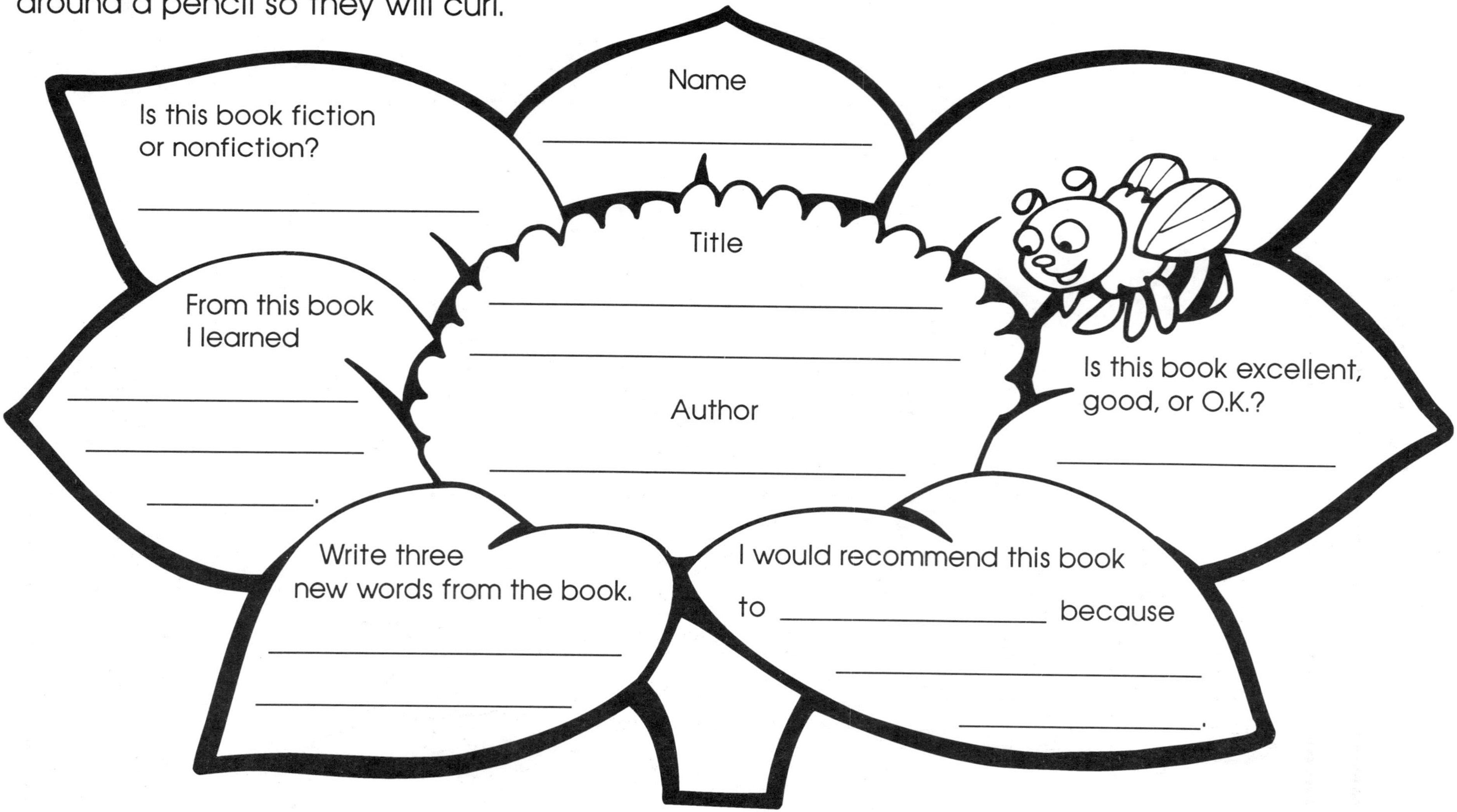

Plant Bookmarks

Seed to Flower

Invite each student to cut brown construction paper to represent soil, and glue it on a large sheet of white construction paper. Give a seed to each student, and have students glue their seed onto the soil part of their paper. Invite students to write and complete the sentence frame *When I plant this seed, I wish it would grow into* ______________. at the top of their paper. Ask students to use craft materials (e.g., construction paper, paint, paintbrushes, wallpaper, fabric, glitter, sequins, pipe cleaners) to make the imaginary plant they wish the seed would grow into. Compile the stories together to make a class book.

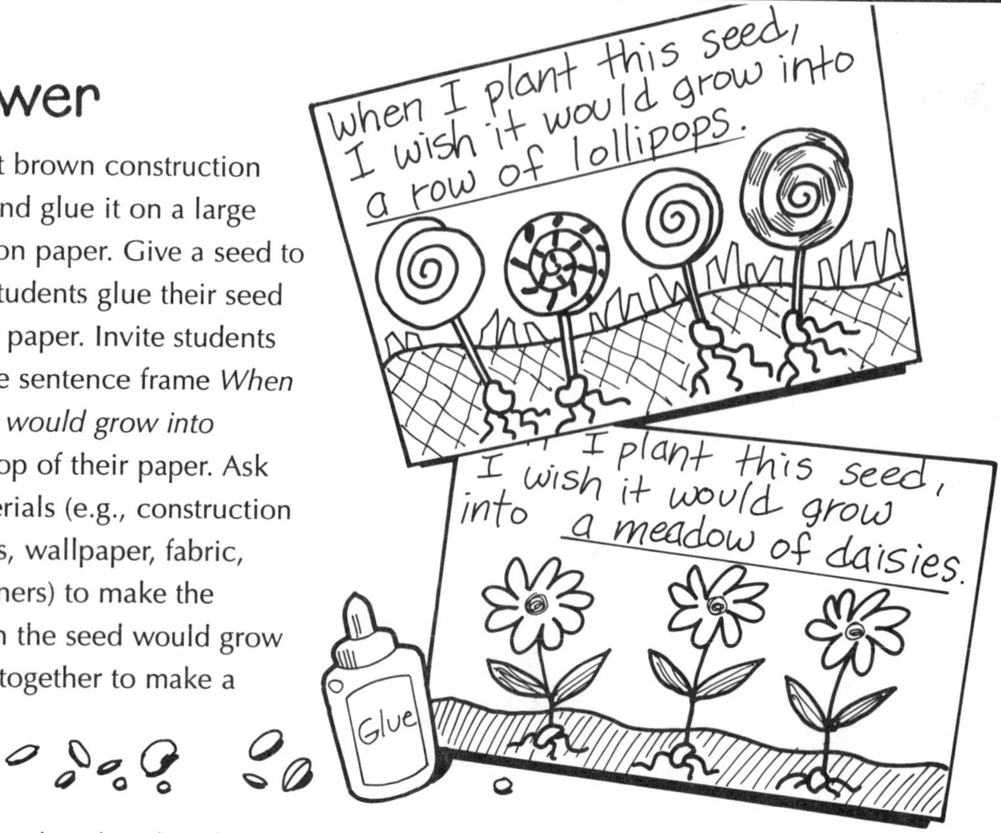

Blooming Personalities

Invite each student to use construction paper and wallpaper scraps to make a flower with a stem, leaves, and petals. Have students glue their photo or a self-portrait in the center of their flower and write a word that describes themselves on each petal. Invite students to share their flower with the class, or display their work as part of a bulletin-board display titled *We're Blooming with Personality.*

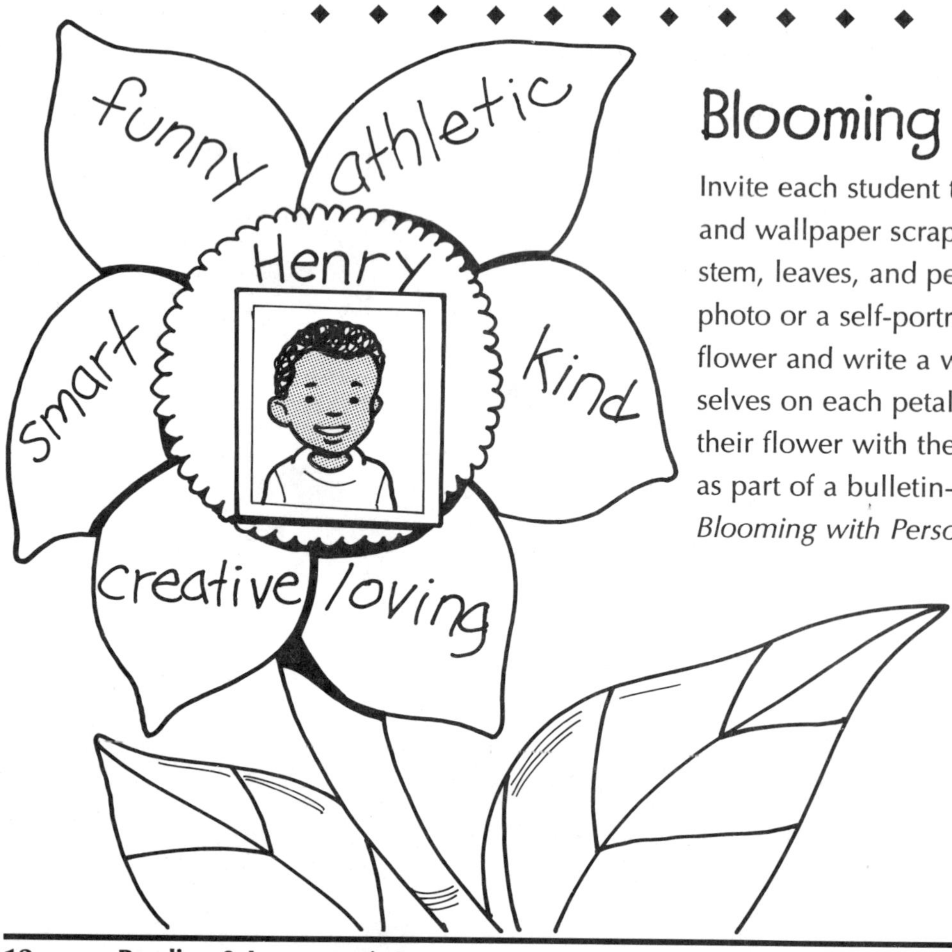

Writing about Plants

Distribute photocopies of the flower-pattern sheet (page 14) for students to use when completing the following writing activities. (You may choose to add writing lines to the sheet before photocopying it for students.)

- Invite students to pretend they are gardeners. Have each student create a "to-do" list for planning and planting a garden. Invite students to illustrate their list.
- Invite each student to write a concrete poem in the shape of a flower. Post student work on a bulletin-board display titled *A Garden of Verse*.
- Read aloud *Barney Bipple's Magic Dandelions* by Carol Chapman. Have students write a story about what they would wish for if they had a magic plant and what might happen if their wish came true. Compile the stories into a book for the class library.
- Have students write stories about a day in the life of a plant. For extra fun, have students draw pictures to match their stories. Invite students to share their stories with the class, or post their work as part of a bulletin-board display.
- Have students list the important things they have learned in the school year. Compile the lists to make a class book, or post their work as part of a bulletin-board display titled *Blossoming with Knowledge*.

Mathematics

Word Problems

Write theme-related word problems on charts or sentence strips to be used at a learning center, or have pairs of students take turns thinking of plant-related word problems for their partners to solve (e.g., *If the gardener picked two roses and four daisies, how many flowers are in his bouquet?*). Encourage students to use seeds to help them solve the problems. To extend learning, invite students to draw pictures to match their word problems.

◆ ◆ ◆ ◆ ◆ ◆ ◆ ◆ ◆ ◆ ◆ ◆ ◆

Seed Sort

Give each student a handful of assorted seeds. Invite students to place their seeds in order from smallest to largest. Have students classify the seeds by different characteristics, such as color, size, or shape. For extra fun, invite students to glue seeds to heavy cardboard to make seed mosaics.

◆ ◆ ◆ ◆ ◆ ◆ ◆ ◆ ◆ ◆ ◆ ◆ ◆

Popcorn-Kernel Estimating

Read aloud *The Popcorn Book* by Tomie dePaola or *Corn—On and Off the Cob* by Allan Fowler. After discussing the story with the class, place popcorn kernels in a jar. Invite students to estimate how many kernels are in the jar. Discuss strategies for making estimations, and record students' guesses on a class chart. Have the class help you count the popcorn kernels, and then compare the total with students' estimations. Pop the kernels and invite students to eat the popcorn for a yummy snack. For extra learning, have students make a class graph of their favorite popcorn flavors.

Sizing Up Plants

Mark the height of each child on a separate large sheet of butcher paper. Invite students to use various art supplies (e.g., paint, paintbrushes, wallpaper scraps, fabric scraps, scissors, glue) to create a plant or flower as tall as their height. Put the finished plants on display in the classroom or school hallway. For extra fun, have students stand next to their life-size plant at the end of the year to see how much they have grown.

◆ ◆ ◆ ◆ ◆ ◆ ◆ ◆ ◆ ◆ ◆ ◆ ◆

Concentrating on Plants

In advance, make two photocopies of the Blooming Math sheet (page 17). On one sheet, write a math problem on each card. On the other sheet, write the answers to those problems on each card. Divide the class into pairs, and give each pair a photocopy of the two sets of cards. Ask students to cut the cards apart, place them facedown, and spread them out on the table with the problems on one side and the answers on the other. Invite partners to take turns choosing two cards, one of each set. If the two cards are a match, the student keeps both cards. If they are not a match, the student places both cards down. The player with the most matches is the winner. To extend learning, invite students to write their own problems and answers on the math cards.

◆ ◆ ◆ ◆ ◆ ◆ ◆ ◆ ◆ ◆ ◆ ◆ ◆

Add a Plant

Help students practice addition skills while reinforcing their knowledge of plants. Draw a large flower on butcher paper, and label the flower parts with numbers to represent points. Tape the flower to the floor. Divide the class into teams of four. Ask each team to think of a plant-related team name. Begin by asking a question related to plants, such as *What part of the flower is the most colorful?* Give the teams several seconds to decide on an answer. Invite each team to reveal their answer. If the answer is correct, invite a student to toss a beanbag on the flower to earn points for the team. Have each team keep track of their points. The team with the most points wins.

Pla

A leaf
makes food
with help from
the sun.

The fruit
carries the
seed of
the plant.

Plants need soil, air, sunshine, and water to grow. Plants grow in all different seasons—even winter.

People use plants in many different ways. Plants can be used to make medicine, lumber, and perfume.

Plants are everywhere. They grow in oceans, mountains, and deserts. They make our world beautiful.

Plant Words

air	living	seed
beautiful	medicine	soil
breathe	mountain	sunshine
desert	ocean	travel
fruit	oxygen	vegetable
grow	season	wind

The stem carries water and minerals from the roots to all parts of the plant.

Illustrated by David Christensen

nts

A flower grows fruit and seeds for new plants

My Plant Book

Name

Plants are living things. There are many kinds of plants. Flowers, grass, and trees are all plants.

Plants are very important. They help give us the oxygen we need to breathe. Plants also give us foods like fruits and vegetables.

Most plants begin as seeds. Seeds can travel to many places. People, animals, even the wind can carry seeds to new places.

The roots
absorb water
and minerals
from the soil.

Blooming Math

Science

Exploring Plants Experiments

Read aloud *The Magic School Bus Plants Seeds* by Joanna Cole. Explain to the class that plants are living things, and that they need food, water, and sunlight to grow. Photocopy the Plants Experiment Record sheet (page 19) for each student, and invite the class to participate in the following experiments. Prior to each experiment, introduce the words *hypothesis, observations,* and *conclusion*. Model how to record the information on the sheet.

Experiment 1: Colorful Petals

Explain to students that the roots and stem of a plant transport minerals from the soil to the parts of a plant above the ground. Ask students to hypothesize how a plant transports water from the base of the plant to the flower. Add several drops of food coloring to water in a clear glass. Trim the stem end of a white carnation, and place it in the dyed water. Leave the flower standing in the water for at least 48 hours. Invite students to observe how the colored water traveled up the stem to the petals. For extra learning, place a carrot in a glass of colored water. After several hours, cut off small sections of the carrot so students can observe how far the dye has been absorbed into the carrot. Invite students to record their observations.

Experiment 2: Far Out Photosynthesis

Explain to students that photosynthesis is the process in which light from the sun mixes with chlorophyll (the green substance in leaves), water, minerals from the soil, and carbon dioxide from the air to create food for the plant. Show students that chlorophyll is produced only in the presence of light. Divide the class into small groups, and give each group a small potted plant with broad leaves and a piece of black construction paper. Invite groups to cut small patterns, such as triangles or squares, out of the construction paper. Have each group use paper clips or tape to attach the shapes to some of their plant's leaves. Ask groups to place their plant in a sunny place for several days. Invite students to hypothesize what will happen to the covered leaves. After several days, ask students to remove the paper from the leaves and look for a color change. The students should notice that the pale areas of the leaves do not contain as much chlorophyll because the black paper prevented photosynthesis from taking place. Invite students to record their observations.

Name ____________________ Date ____________________

Plants Experiment Record

Experiment Name ____________________

My hypothesis: I predict ____________________

____

Before	After
Date: ____________________	Date: ____________________
Observations: ____________________	Observations: ____________________
Illustration	Illustration

My conclusion: I learned ____________________

____

Peek-and-See

In advance, empty an entire bag of lima beans in a bowl of water and soak them overnight. Read aloud *One Bean* by Anne F. Rockwell. After discussing the book with the class, give each student a large sheet of green construction paper. Invite students to fold the sheet of paper in half. Ask students to draw a large bean on one half of their paper. Have students cut out their large bean shape, so there are two bean shapes. Ask students to write *What's Inside a Seed? Peek-and-See* on one of their bean shapes. Give each student a soaked lima bean. Invite students to remove the seed coat and pull the bean apart. Invite students to use construction paper, crayons, and glue to make the tiny plant inside the seed on the other bean shape. Have students staple the two bean shapes together to make a small book. For extra learning, ask students to bring in other seeds, such as peanuts, almonds, or sunflower seeds, to soak, open, examine, and draw. Have students discuss and compare the different seeds.

Growing Mobiles

Read aloud *From Seed to Plant* by Gail Gibbons. Invite students to make mobiles illustrating the growth of a plant. Distribute index cards and crayons or markers to each student. Invite students to illustrate on separate index cards the different stages of plant growth from seed to plant. For an extra challenge, have students write about each stage. Have students hole-punch the top and bottom of each card and thread string through each hole. Hang the mobiles from the classroom ceiling.

Hungry for Plants

Read aloud *Growing Vegetable Soup* by Lois Ehlert. Discuss with students how we eat many different parts of plants: leaves, stems, flowers, roots, fruits, and seeds. Take students on a field trip to a grocery store, or have them imagine they are at one. Ask students to brainstorm foods we eat from plant parts, such as carrots (roots), spinach (leaves), and broccoli (flowers). Record their responses on chart paper or the chalkboard. Give each student a copy of the Plants We Eat! sheet (page 22). Invite students to glue or draw pictures of plant parts we eat in the appropriate categories. Have each student choose a vegetable from the sheet to bring to school. Make a vegetable soup by having students wash and chop their vegetable (with adult supervision) before placing it in a soup pot with 8 cups (2 l) of water, 8 bouillon cubes, and rice or pasta (optional). Simmer for 30 minutes. Add salt and pepper and serve!

◆ ◆ ◆ ◆ ◆ ◆ ◆ ◆ ◆ ◆ ◆ ◆ ◆

Parts of a Flower

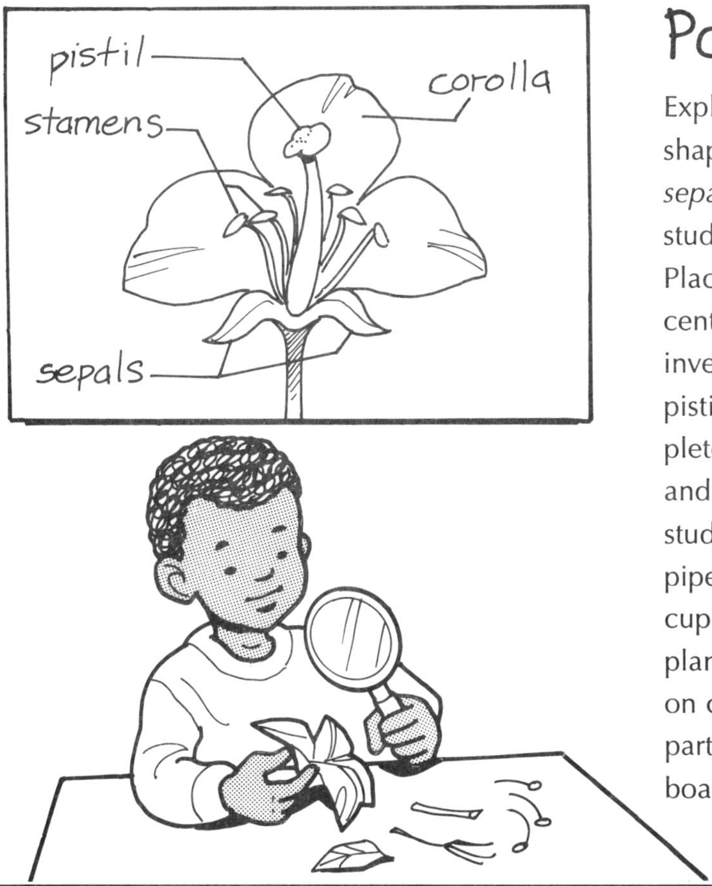

Explain to students that flowers come in all shapes, sizes, and colors. Introduce the terms *sepals, corolla, stamens,* and *pistils.* Give each student the Parts of a Flower sheet (page 23). Place several kinds of flowers in a learning center, and ask students to use a hand lens to investigate the number of petals, stamens, and pistils on each flower. Invite students to complete the flower sheet as they learn about flowers and their parts. To extend learning, invite each student to use art materials (e.g., tissue paper, pipe cleaners, paper scraps, wallpaper samples, cupcake liners) to design one large flowering plant (with all of the parts) and glue the plant on construction paper. Have students label each part. Display the plants as part of a bulletin-board display.

Name ______________________ Date ______________________

Plants We Eat!

Directions: Glue or draw pictures of plant parts we eat in the boxes below.

Leaves	Stems	Flowers	Roots	Fruits	Seeds

What plant do you like eating most? ______________________

Name ______________________ Date ______________________

Parts of a Flower

Directions: Read about the parts of a flower. Then use the bolded words to label the parts of the flower.

1. The **corolla** is the part of the flower with petals. It attracts birds and insects that pollinate the flower and help it grow.
2. Most flowers have several **stamens.** They contain the powdery part of the plant called pollen.
3. **Sepals** are the leaf-like parts that protect the flower bud.
4. The **pistil** is in the center of the flower. It has a sticky surface to attract pollen.

Social Studies & Geography

Flowers across America

Explain to students that each state has its own flower. Invite each student to choose a state and draw the state's flower on a sheet of paper. Post a United States map, and have students use yarn pieces to pin their flower drawing to the appropriate place on the map. For an extra challenge, have students research information on a flower and write down any facts they learned.

◆ ◆ ◆ ◆ ◆ ◆ ◆ ◆ ◆ ◆ ◆ ◆ ◆

A More Beautiful Place

Read aloud *Miss Rumphius* by Barbara Cooney. Discuss with students how Miss Rumphius made her environment more beautiful by planting flowers. Invite students to brainstorm ways they would make the world a more beautiful place. Record their responses on chart paper. Invite each student to write a letter that thanks Miss Rumphius for making the world a more beautiful place and explains how he or she plans to make the world more beautiful. Have students share their letter with classmates, or present their work as part of a bulletin-board display.

◆ ◆ ◆ ◆ ◆ ◆ ◆ ◆ ◆ ◆ ◆ ◆ ◆

Growing with Responsibility

Ask students to imagine that their life is a garden, and discuss the responsibility it takes to tend a flower garden. Cover a bulletin board with light-blue butcher paper. Use a permanent marker to write each student's name on a bean-shaped piece of green construction paper. When students show responsibility, have them glue their bean near the bottom of the paper. Give each student three strands of brown yarn for roots and a construction-paper stem and flower. Invite students to add one item to the board each time they perform a responsible act. Soon the wall will turn into a "garden of responsibility."

Musical Plants

In advance, buy green crepe paper streamers. Give each student a streamer. Invite students to curl their body on the ground as if they were a seed in soil and hold their streamer in a ball. Play gentle music, and have students pretend they are a seed slowly beginning to grow. Ask students to make motions that would show the plant being watered, getting sunshine, and responding to a gust of wind. Encourage students to use their streamer as if it were the parts of a plant beginning to grow. Invite students to take turns calling out directions for the growing "plants."

Garden Gossip

In advance, draw a simple garden scene (e.g., flowers, sun, scarecrow, animal and insect guests) on the chalkboard or use the Garden Time mural (page 4). Ask students to imagine and discuss what the flowers, scarecrow, sun, and garden friends might be thinking and saying to each other. Record their responses on the chalkboard or chart paper. Divide the class into small groups of four or five students. Ask students to use art supplies (e.g., construction paper, fabric, markers, paint, glitter, feathers) to make props and costumes. Have each group write a scene with dialogue to go with the mural. Invite each group to dramatize their garden scene for the class.

Physical Education

Skip to My Garden

Invite students to chant the following rhyme as they jump rope. For extra fun, encourage students to make up more verses.

I went to the garden to pick some flowers,
But it took me several hours.
How many hours did it take?

I went to the garden to pull some weeds.
Instead I ate sunflower seeds.
How many seeds did I eat?

Garden Relay

Set up a relay course in the classroom or on a playfield. Place three chairs in a row approximately 3' (1 m) apart. Place a bean-filled pot on each chair. Place three more chairs approximately 10' (3 m) away from the first row, and set an empty flowerpot or planter on each one. Divide the class into three teams, and have each team line up behind one of the bean-filled pots. Give the first player on each team a spoon. At the starting signal, have the first players scoop up as many beans as possible, carry them to the empty flowerpot or planter, and then give the spoon to the next team member. If a student drops any beans, he or she must pick them up and return to the starting point. The team that empties their pot first is the winner.

Arts & Crafts

Pressed Flowers

Invite students to collect an assortment of leaves and small flowers. Have each student place the leaves and flowers inside a folded sheet of paper and lay a heavy book on top of it for one or two days until the leaves and flowers are flat. Distribute two sheets of waxed paper to each student. Ask students to arrange their leaves and flowers between the two pieces of waxed paper. Cover the waxed paper with a thin cloth, and press it firmly with a warm iron. (Only adults should handle the iron.) Have students glue their waxed paper to a construction-paper frame, and share student work as part of an art display. To extend learning, invite students to write a poem to go with their pressed flowers.

Petal Prints

In advance, cover student work areas with newspaper. Provide each student with cups of tempera paint (assorted colors) or ink pads, white construction paper, and cotton swabs. Invite students to use their fingerprints to paint or stamp a pattern resembling plants and flowers. Have students use cotton swabs to paint green stems. Glue each picture to a construction-paper frame, and share student work as a bulletin-board display. For extra fun, have students create wrapping paper or stationery with plant and flower designs.

Crystal Gardens

In advance, cover student work areas with newspaper. Fill small paper cups with tempera paint (assorted colors) for each student. Dissolve 2 tablespoons (30 ml) of salt in each cup to make "salt paint." Give each student drawing paper, salt paint, and paintbrushes. Invite students to paint garden scenes with the salt paint. When the paint is dry beautiful salt crystals will appear to give a glistening effect. Glue student work to colorful construction paper, and present it as part of a bulletin-board display.

Beautiful Blooms

In advance, cover student work areas with newspaper. Give each student a coffee filter, a paintbrush, a cup of water, and watercolor paints. Invite students to flatten their filter, paint it with water, and then dot it with a few drops of paint. The paint will spread into designs as it is absorbed by the filter. Allow the filters to dry completely, and then have students attach a green construction-paper stem and leaves to it. Display the "blooms" in classroom windows to give them a transparent effect.

◆ ◆ ◆ ◆ ◆ ◆ ◆ ◆ ◆ ◆ ◆ ◆ ◆

Garden Scenes

Invite students to read any plant-related books (see Recommended Books, page 32). Ask each student to use art materials (e.g., clay, paint, paintbrushes, construction paper, glue, fabric, pipe cleaners) to make a shoe-box diorama of a plant or garden environment. For extra learning, have students write stories or interesting plant facts to go with their diorama. Invite students to display their creation for parents during open house.

◆ ◆ ◆ ◆ ◆ ◆ ◆ ◆ ◆ ◆ ◆ ◆ ◆

Pleasant Potpourri

In advance, ask each student to bring several flower blossoms from home or a florist. Have students pull apart their blossoms and place them on baking sheets to air dry for several days. When the petals are completely dry, place them in a large mixing bowl. Invite students to sprinkle the petals with spoonfuls of mixed spices, such as rosemary, cinnamon, and allspice. Add a few drops of potpourri oil. Have students mix the petals and spices. Place the potpourri in small containers around the classroom to freshen the air. For extra fun, invite students to make sachets. Have each student place 2 tablespoons (30 ml) of dry potpourri in the center of a fabric square, secure the fabric with a rubber band, and tie a ribbon around it.

Chocolate Dirt

Supplies:

- 2 packages of chocolate sandwich cookies
- rolling pin, mixing bowls, and spoon
- 3 boxes instant chocolate pudding
- 6 cups milk
- new, large flowerpot (no bottom hole)
- silk, plastic, or fresh flowers
- gummy worms
- new, clean trowel
- paper plates
- plastic spoons

Crush the cookies with a rolling pin to resemble potting soil. In a large mixing bowl, combine pudding and milk according to the directions on the pudding box. Put some of the crushed cookies in the bottom of a flowerpot. Use a spoon to add half the pudding mixture, then add another layer of cookies. Continue to add layers of pudding and crushed cookies, ending with a layer of cookies. Refrigerate for one hour. Before serving, place flowers and gummy worms in the chocolate "dirt." Dig out portions of the dirt with a trowel, and serve it on paper plates. Invite students to eat their yummy treat.

◆ ◆ ◆ ◆ ◆ ◆ ◆ ◆ ◆ ◆ ◆ ◆ ◆

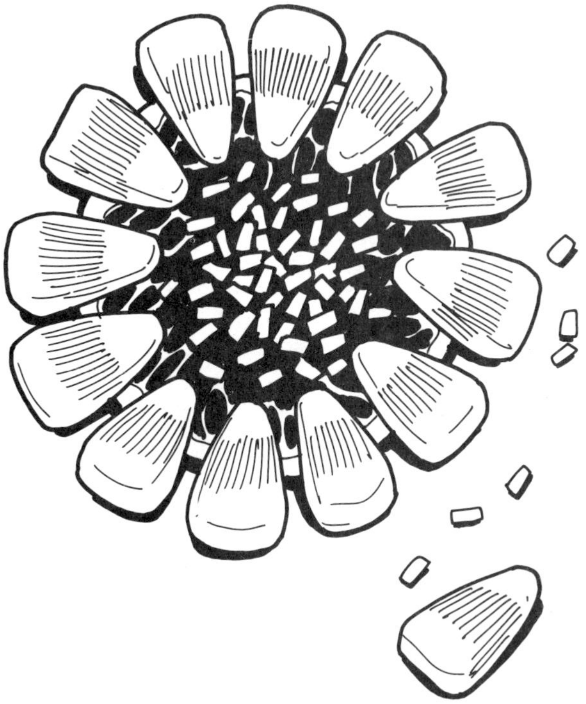

Sunshine Cookies

Supplies:

- round sugar cookies
- plastic knives
- chocolate frosting
- candy corn
- yellow sprinkles

Have each student use a plastic knife to spread chocolate frosting on a sugar cookie and add candy corn "petals." Have students add yellow sprinkles as the "seeds" of their sunflower. For extra fun, invite students to eat their cookie flower while you read aloud *The Tiny Seed* by Eric Carle or *From Seed to Sunflower* by Gerald Legg.

Special Events

Field Trip

Take your class to a nature center or garden to examine plants and flowers. For extra learning, take the class on a walk of the school grounds or a local park to observe and compare plants and flowers in their natural habitat. Invite a gardener or botanist who specializes in the study of plants to tell about the plant species the class identifies.

◆ ◆ ◆ ◆ ◆ ◆ ◆ ◆ ◆ ◆ ◆ ◆ ◆

Garden Tours

Discuss with students the duties of a tour guide on a nature walk. Challenge students to create classroom exhibits, charts, and posters of flowers and plants. Invite students to be "tour guides" for classroom exhibits and presentations. Invite other classrooms, parents, and guests to take a tour of the classroom as the tour guides present their knowledge of plants.

◆ ◆ ◆ ◆ ◆ ◆ ◆ ◆ ◆ ◆ ◆ ◆ ◆

Flower Festival

Invite students and parents to a "Flower Festival." Decorate the classroom with student-made plant and flower artwork (see Arts & Crafts, pages 27–28). Ask parents to prepare a "Plants Potluck" of foods made from fruits and vegetables. Invite students to offer nature walks through the classroom, sing songs, and read or present creative writing projects to share what they have learned about plants. Be sure to take pictures of this special event to add to the class scrapbook!

◆ ◆ ◆ ◆ ◆ ◆ ◆ ◆ ◆ ◆ ◆ ◆ ◆

Garden Careers

Invite people in plant professions, such as botanists, forest rangers, farmers, landscape gardeners, and florists, to your classroom for a "Career Day." Encourage students to listen to the speakers and write thank-you letters sharing what they learned about each profession.

You Are a Blooming Success

____

has successfully completed
the Plants Unit!

Signed ____________________ Date ____________________

A Big Bouquet of Thanks to

____

for ____________________________

Signed ____________________ Date ____________________

Recommended Books

Alison's Zinnia by Anita Lobel
Anna's Garden Songs by Mary Q. Steele
An Apple Tree Through the Year by Claudia Schnieper
Barney Bipple's Magic Dandelions by Carol Chapman
Bean and Plant by Christine Back
Berries, Nuts, and Seeds by Diane L. Burns
Cactus by Cynthia Overbeck
Carnivorous Plants by Nancy J. Nielsen
The Carrot Seed by Ruth Krauss
Corn Belt Harvest by Raymond Bial
Corn—On and Off the Cob by Allan Fowler
Desert Giant: The World of the Saguaro Cactus by Barbara Bash
Exploring an Ocean Tide Pool by Jeanne Bendick
The Flower Alphabet Book by Jerry Pallotta
From Seed to Plant by Gail Gibbons
From Seed to Sunflower by Gerald Legg
Growing Vegetable Soup by Lois Ehlert
Jack and the Beanstalk by Matt Faulkner
The Legend of the Indian Paintbrush by Tomie dePaola
Linnea in Monet's Garden by Christina Bjork
Linnea's Windowsill Garden by Christina Bjork
The Magic School Bus Gets Planted: A Book about Photosynthesis by Joanna Cole
The Magic School Bus Plants Seeds by Joanna Cole
Miss Rumphius by Barbara Cooney
A Nice Walk in the Jungle by Nad Bodsworth
One Bean by Anne F. Rockwell
Over Under in the Garden: An Alphabet Book by Pat Schories
Planting a Rainbow by Lois Ehlert
Plants that Never Ever Bloom by Ruth Heller
The Popcorn Book by Tomie dePaola
Popcorn Plants by Kathleen V. Kudlinski
The Reason for a Flower by Ruth Heller
The Red Poppy by Irmgard Lucht
The Tiny Seed by Eric Carle
Tomato by Barrie Watts
A Walk in the Rainforest by Kristin Joy Pratt
What Makes a Flower Grow? by Susan Mayes